To the state of
Florida

♡
Sandra Markle
2024

WILD INVENTIONS

Ideas Inspired by Animals

Sandra Markle

M Millbrook Press / Minneapolis

For Adrianne Elkins and all the children of Coyote Willow Family School in Albuquerque, New Mexico

Acknowledgments: The author would like to thank the following people for sharing their enthusiasm and expertise: Dr. Kollbe Ahn, Marine Science Institute, University of California, Santa Barbara, Santa Barbara, California; Dr. Brian Branstetter, National Marine Mammal Foundation, San Diego, California; Dr. Alfred Crosby, University of Massachusetts Amherst, Amherst, Massachusetts; Dr. Kevin Foster, University of Oxford, Oxford, United Kingdom; Dr. Elliot Hawkes, University of California, Santa Barbara, Santa Barbara, California; Dr. Laura Kloepper, University of New Hampshire in Durham, New Hampshire; Dr. Samuel Ocko, Stanford University, Stanford, California; Mick Pearce, architect, Avondale, Harare, Zimbabwe; Sir John Pendry, Imperial College London, London, United Kingdom; Dr. Jeff Pettis, Apimondia, International Federation of Beekeepers' Associations, Rome, Italy; Dr. Simon Pollard, University of Canterbury, Christchurch, New Zealand; Dr. Chris Roh, California Institute of Technology, Pasadena, California; Dr. Alexandra Schnell, University of Cambridge, Cambridge, United Kingdom; Dr. Russell Stewart, University of Utah, Salt Lake City, Utah; Dr. Edwin L. Thomas, Texas A&M University, College Station Texas; Dr. Gabriel D. Weymouth, Southampton Marine and Maritime Institute, University of Southampton, Southampton, United Kingdom.

A special thank-you to Skip Jeffery for his loving support during the creative process.

Millbrook Press™
An imprint of Lerner Publishing Group, Inc.
241 First Avenue North
Minneapolis, MN 55401 USA

For reading levels and more information, look up this title at www.lernerbooks.com.

Illustrations on pages 12, 22, 31, 44 by Laura K. Westlund.

Designed by Kimberly Morales.

Main body text set in Johnston ITC Std.
Typeface provided by International Typeface Corporation.

Library of Congress Cataloging-in-Publication Data

Names: Markle, Sandra, author.
Title: Wild inventions: ideas inspired by animals / Sandra Markle.
Description: Minneapolis: Millbrook Press, [2024] | Series: Sandra Markle's science discoveries | Includes bibliographical references and index. | Audience: Ages 9–12 | Audience: Grades 4–6 | Summary: "Where do inventors find inspiration? Sometimes they look to animals! Author Sandra Markle presents twelve animal adaptations and the amazing inventions they've inspired"—Provided by publisher.
Identifiers: LCCN 2022052151 (print) | LCCN 2022052152 (ebook) | ISBN 9781728467955 (lib. bdg.) | ISBN 9798765602270 (eb pdf)
Subjects: LCSH: Inventions—Juvenile literature. | Biomimicry—Juvenile literature. | Animals—Miscellanea—Juvenile literature. | Adaptation (Biology)—Miscellanea—Juvenile literature.
Classification: LCC T48 .M368 2024 (print) | LCC T48 (ebook) | DDC 600—dc23/eng/20221110

LC record available at https://lccn.loc.gov/2022052151
LC ebook record available at https://lccn.loc.gov/2022052152

Manufactured in the United States of America
1-51836-50463-3/30/2023

CONTENTS

GOING WILD!

Since ancient times, people have observed the special physical features and behaviors—called adaptations—that enable wild animals to survive and even thrive. When animals give people ideas for inventions, it's called *bioinspiration*.

For example, some animals need to camouflage, or blend into their home environment. Camouflage allows arctic foxes to ambush the small animals they eat and avoid being seen by bigger animals that could attack them. And on the walls of caves in the Pamir Mountains of central Asia, colored-earth sketches show human hunters wearing animal skins to camouflage themselves. Researchers estimate this cave art was drawn around eight thousand years ago. These ancient people probably were bioinspired to camouflage themselves this way, though no written record confirms this. But they would not have used the term *bioinspired*. That concept wasn't given a name until the 1960s. And it wasn't until the 1990s that people started to seek bioinspiration when they needed to design an invention to tackle a problem.

An arctic fox has a brown summer coat, which it begins to shed in September. By November it has a snow-white, thick, wooly winter coat. The coat keeps it warm until May, when the fox starts to shed and regrow its brown summer coat.

Giant cuttlefish manage a quick-as-a-blink color change by controlling their three layers of skin cells. The cells contain color, and altering their size changes the cuttlefish's appearance.

Giant cuttlefish also camouflage, which helps them blend into their coral reef home. And when a predator, such as a shark, swims close to a giant cuttlefish, it must camouflage quickly to hide. In the early twenty-first century, Edwin L. Thomas reported that he was amazed and bioinspired by the giant cuttlefish's color-changing ability. So, he led his research team at Rice University in Texas in experimenting on how to make colors change quickly on a screen. Thomas wanted to invent a new kind of display screen, possibly a future television screen, made from layers of hard glass and soft, swellable gel. Mimicking how the swelling and shrinking in a cuttlefish's skin displays colors, the team chemically treated the gel layers to control the amount of swelling in different parts of the layered stack. The thickness of the gel controlled the wavelength of light being reflected. And that determined which color displayed on the screen. While it worked, construction of the screen was too expensive and this invention was never produced for people to use. But the way animals—such as cuttlefish—create colors for camouflage continues to be a hot research topic for inventors.

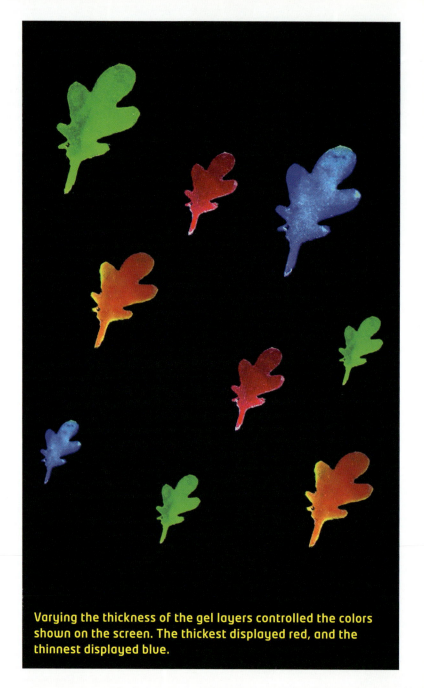

Varying the thickness of the gel layers controlled the colors shown on the screen. The thickest displayed red, and the thinnest displayed blue.

An arctic fox's summer coat is camouflaged brown, which helps it blend into its environment and ambush prey.

As you just discovered, more than one animal can adapt to meet the same life challenge. But which adaptation is better bioinspiration for an invention? Or do both inspire inventions for different reasons?

Turn the page to keep exploring bioinspired inventions.

And get ready to be AMAZED!

In addition to changing color, cuttlefish can retract or extend bumpy spots on their skin to perfect their camouflage cover when hiding from predators.

COOL HOUSE RULES

Honeybees and mound-building termites build very different homes, but both face the same challenge: *How to build a home that doesn't get too warm when the weather heats up.*

Honeybees and termites are social animals—they live in large groups. Their homes must stay at a safe temperature for both the adults and the more temperature-sensitive developing young, called larvae.

Honeybees build hives for homes using wax they produce themselves. The workers in the colony perform different tasks during different stages of their six- to eight-week life. At about three weeks old, workers produce wax flakes that emerge between the segments of their abdomens. The workers use these flakes to build new wax cells and expand the hive or to repair existing cells in the hive. The cells are where the queen deposits her eggs and where the larvae complete their development after hatching. The bees also store the colony's food supply in cells within the hive.

A mound-building termite colony's adult population includes reproducing adults (usually a single male and female, called the king and queen), soldiers with large heads and mandibles to defend the colony, and wingless, soft-bodied workers that build the mud walls, collect food, and care for the young.

A honeybee colony includes one reproductive female (called the queen), a small number of males called drones (mates for new queens), and thousands of female bees that do not reproduce called workers. The workers have different jobs at different ages to clean the hive, feed the young, build or repair the hive, and collect nectar to produce honey.

This wild honeybee hive is in a tree, and the sections of honeycomb are clearly visible. Spaces in between the sections add airflow and aid the worker bees' efforts to cool the hive.

When summer temperatures heat up, honeybees team up to cool their hive. Some workers fly off to drink a bellyful of water at the closest puddle or other water source. Back home, they spit the water onto any exposed parts of the hive. Then other bees join them in fanning their wings to increase airflow in the hive. The bees seem to naturally know that water evaporating (turning from liquid water to water vapor, a gas) has a cooling effect. That happens because when water evaporates, a little heat energy is used up. This takes heat away from the hive, leaving a cooler environment. If you've ever felt cool after taking a bath or shower, you know this works.

Mound-building termites use a different system to keep their homes cool. Samuel Ocko, a scientist at Stanford University in California, said, "Without a blueprint or building supervisor, millions of blind, rice grain-sized termites work together to construct a towering mound around a large central chimney. That gives their underground nest a complex ventilation system." Termite workers build their underground nest little by little as each insect mixes its saliva with dirt to make a dab of mud. Then one dab at a time, the termites build walls, leaving spaces between them to create chambers and passageways. The extra dirt is carried up and out of the nest to form the mound. The termites also build more passageways in the mound. These passageways act like air ducts, directing airflow within the mound. Researchers don't yet know how termite workers sense when interior walls need to be remodeled to redirect passageways and change the airflow through their home, but they do it!

Mound-building termite mounds average 9 feet (2.7 m) tall, but some found in parts of Australia are as much as 26 feet (8 m) tall!

Ocko explained, "By constantly digging open new vent holes [in the passageways], plugging old ones, and redirecting channels, the termite workers regulate air currents through the mound." During the day, hot air flows up the outside of the mound, causing cooler air to flow down into the nest. At night, cool air flows down the outside of the mound and warm air rises through the nest up the chimney. Though the mound passageways are not open to the outside, the mound acts like an air conditioner as stale air leaks out and fresh air seeps in.

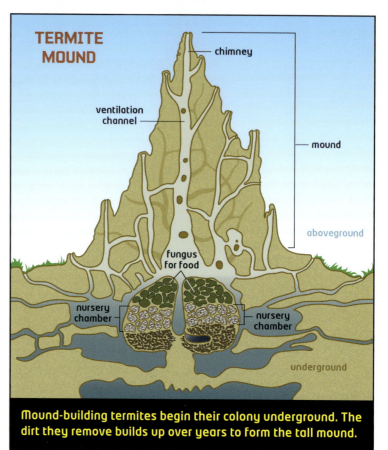

TERMITE MOUND

chimney

ventilation channel

mound

aboveground

fungus for food

nursery chamber

nursery chamber

underground

Mound-building termites begin their colony underground. The dirt they remove builds up over years to form the tall mound.

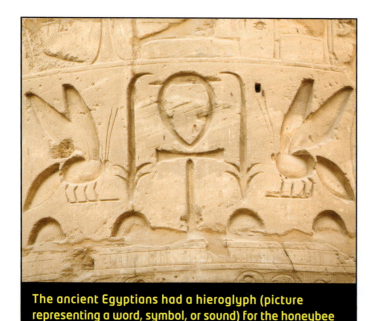

The ancient Egyptians had a hieroglyph (picture representing a word, symbol, or sound) for the honeybee because honeybees were so prized for the honey they produced.

So, how have these two different animals sparked bioinspiration?

Paintings on the walls of an Egyptian pharaoh's tomb show enslaved people fanning the air over water-filled pots. While possibly someone in ancient Egypt thought up this idea, it's just as likely this inventive way of cooling air was inspired by observing honeybees.

No one knows for certain how swamp coolers got their name. Some say it's because the damp material inside could sometimes look and smell like a swamp.

Once an invention is successful, its use spreads and creative people modify it—even use new technologies to improve how it works. So, early settlers in the American Southwest used evaporative cooling that worked without manual labor. They simply hung wet cloths over open windows and doorways.

The modern version of this early invention for cooling air inside a building is called an evaporative cooler. In these devices, electric fans push hot, dry, outside air through water-soaked pads of material, such as wood fibers or a plastic mesh. The hot air evaporates the water in the pads and cools before entering the building. A building cooled with a swamp cooler, as evaporative coolers were sometimes called, is only ever a little cooler than the outside air. But this method of cooling uses much less electricity than most of today's air-conditioning systems.

Architect Mick Pearce reported that he was inspired by the mound-building termites' method for creating airflow. And it influenced his design for the Eastgate Centre, a shopping and office building in Harare, Zimbabwe. Eastgate has big chimneys that act like a termite mound's chimney towers. But how this building is cooled is a little different from how the termite's home stays cool.

At night, vents in the lower part of the building open to let cooler outside air flow in. Fans on the first floor of the building help push the air upward. The warmer air inside the building rises and exits through the chimneys. During the day, the vents close, holding in the cooler air. The building's thick walls act like superinsulation by absorbing much of the outside heat, blocking it from affecting the interior environment. By nighttime the warmed-up exterior walls radiate enough heat to increase the building's interior temperature. Then lower vents are opened, repeating the cooling process. This termite-inspired building design keeps Eastgate's interior comfortable year-round at about a 35 percent lower energy cost than buildings of similar size in Zimbabwe that have typical commercial air-conditioning systems.

Irregular projections on the outside of Eastgate Centre reduce the amount of direct sunlight on the walls, limiting heat absorption. Like the termite mound's mud, Eastgate's brick-and-concrete exterior walls soak up the heat, which helps keep the interior cool.

Large open spaces inside Eastgate Centre help the building mimic the self-cooling termite mound's central chimney.

SEE WHAT I HEAR

A dolphin swims in the ocean's depths. A bat flies through the air at night. Despite their very different environments, these animals face the same challenge: *How to safely travel when they can't see what's around them.*

Both animals adapted to tackle this challenge by developing a way to blast out sounds and analyze the echoes that bounce back. By doing this, each animal becomes aware of its surroundings even if it can't use its eyes to see what is there. The ability to sense the environment this way is called *echolocation.*

While using echolocation, a dolphin can produce as many as one thousand sounds per second.

The horseshoe bat uses echolocation to find and catch insects to eat when it's too dark to see clearly.

Brian Branstetter, a researcher at the National Marine Mammal Foundation in California, said, "Dolphins can sense something golf ball–sized about 240 feet [73 m] away and home in on it even in dark, murky water. They can sense something as big as a submarine from nearly a mile [1.6 km] away." A dolphin produces click sounds by inhaling air through its blowhole (breathing hole on top of its head) and pushing the air back and forth between its lungs and the air sacs inside its head. Next, it projects the clicks forward through a fatty part inside its forehead called the melon. That amplifies the clicks and blasts them out. These clicks bounce back off anything they strike. The dolphin picks up the bounced-back echoes as vibrations within its jawbone. These vibrations travel to its ears, which send signals to its brain. Once its brain interprets these signals, the dolphin senses whatever bounced back its clicks.

These dolphins off the coast of South Africa use echolocation to find prey, such as a school of sardines.

Even when humans are limited to using just our sense of touch, we can still figure out a lot about an object.

waves penetrate something before bouncing back.

Laura Kloepper, a bat researcher at the University of New Hampshire in Durham, New Hampshire, reported, "Dolphins can echolocate farther than bats because sound travels nearly four times faster through water than through air. But bats can sense something as small as a moth in the dark from about fifteen feet [4.5 m] away. Then [they] home in and catch it."

To echolocate, bats also make click sounds. Some do this with wing snaps, but most use their tongues. Then sounds exit through the bat's mouth or nose. And when a bat detects the echoes, its brain analyzes those signals. Like a dolphin, a bat senses the shapes of whatever is around it. But bats can do something dolphins can't. Individual bats can vary the pitch and frequency of their clicks to be unique.

Producing its own special echolocation sounds is important when a bat is flying in a large group. While dolphins usually travel in groups of just five to twenty individuals, bats often fly in crowds of a thousand or more. And each bat in this giant

Brian Branstetter explained, "What a dolphin senses is like you reaching into a box to feel what's inside [without looking]. You sense what's there, but you don't get all the details you would with your eyes." Through echolocation a dolphin gets a three-dimensional sense of the shapes of whatever is in its environment. But in addition to an object's shape, a dolphin also senses its thickness. That would be like a person being able to touch a melon and sense how thick the rind is. A dolphin's echolocation lets it gain this additional information because sound

Laura Kloepper holds Belle, a Harris's hawk ready to be released and fly among the swarming bats. The tiny camera and microphone were designed to be small and light enough for Belle to wear on her head without affecting her flight.

behavior and echolocation sounds, Kloepper outfitted Belle with a custom-designed camera and microphone powered by a battery backpack. She said, "What we discovered was *amazing!* We learned each bat is able to change the pitch and duration of its clicks to be unique. That way, it isn't confused by any other bat's sounds or echoes."

So, how have these two different animals sparked bioinspiration?

Ships needed to sense and avoid unseen, underwater hazards, such as part of an iceberg. The 1912 sinking of the *Titanic* when it struck an iceberg motivated British scientist Robert W. Boyle to invent a device to help prevent such future disasters. In 1917 he shared his invention, a device that pulsed out sounds called *pings* underwater and recorded the bounced-back echoes. The device displayed those echoes as electronic dots on a screen, creating a hint of whatever was below the surface. Was Boyle's invention inspired by dolphins echolocating?

group is likely to be zooming through the air at 60 miles (96 km) per hour. So, it's helpful that each bat can both produce its own unique echolocation sounds and quickly recognize them. That lets each individual bat navigate safely.

Kloepper learned about bats making signature echolocation sounds when she sent a spy on a series of flights among groups of bats in New Mexico. The spy was Belle, a Harris's hawk trained for its test flights by Kloepper's research team member Paul Domski. To record the bats' in-flight

Millions of Mexican free-tailed bats swarm out of a cave in Carlsbad, New Mexico. They all sleep in the cave during the day and leave as it gets dark to catch and eat flying insects.

He didn't leave a written record identifying the inspiration for his invention. But how it works remarkably mimics dolphin echolocation. During the 1930s, Americans began referring to this invention as *sonar* (sound navigation ranging), and the name caught on. During World War II, sonar became extremely important as sea battles were often between surface ships and submarines.

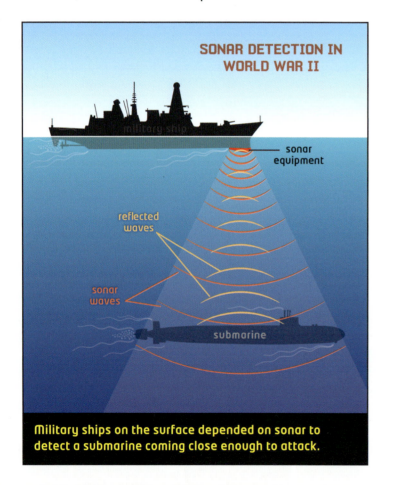

SONAR DETECTION IN WORLD WAR II

sonar equipment

reflected waves

sonar waves

submarine

Military ships on the surface depended on sonar to detect a submarine coming close enough to attack.

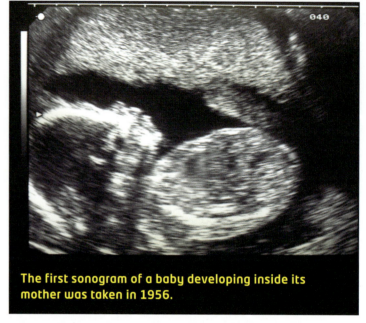

The first sonogram of a baby developing inside its mother was taken in 1956.

Like a dolphin's echolocation, sonar *pings* were also able to penetrate far enough to provide a sense of the thickness of whatever they bounced off. Researchers figured out how to project sound waves into the human body and collect echoes to create images of internal parts. Such an image is called a sonogram. Today, 3D ultrasound imaging directs high-frequency sound waves into the body from different angles. Then receiving equipment collects the echoes and a computer transforms those signals into a three-dimensional image.

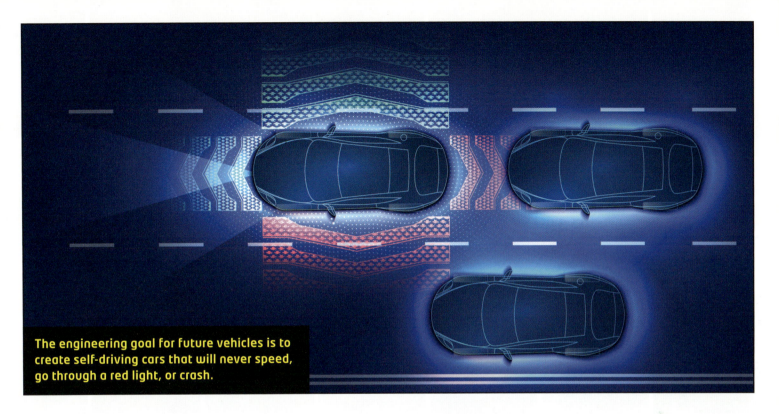

The engineering goal for future vehicles is to create self-driving cars that will never speed, go through a red light, or crash.

Engineers developing self–driving cars are inspired by how bats use personalized echolocation to avoid collisions. These cars use lidar (light detection and ranging) technology instead of sonar as they navigate because light travels faster than sound. But the process is the same. Each second, the lidar equipment shoots out laser pulses that are harmless and invisible to the human eye. These pulses bounce off everything they touch. The car's equipment picks up those reflections and records the length of time it takes for them to return. That indicates the distance from the car to what bounced it back. Any pulse bouncing off an obstacle that is too close activates the car's automatic braking system. This technology can also keep a car in its lane on a road while the onboard Global Positioning System (GPS) directs it from one location to another. Someday, each car's lidar signal could be unique. Then a car could transport passengers without a driver and avoid crashes even in heavy traffic!

STICK TO IT

To thrive where they live, geckos and sandcastle worms face the same challenge: *How to stick things together.* Both animals adapted to overcome this challenge, but they did so in very different ways. That's partly because these animals have different goals. Geckos need to be able to quickly stick and unstick themselves to something, but sandcastle worms need to stick materials together and have them stay together—even underwater.

If you live in an area where there are geckos, you've probably watched those little lizards do something seemingly impossible. They can run up a vertical wall. Their toes create an adhesive force—a force that holds two materials together at their surfaces. But geckos don't produce any gluelike substance on their toes. A gecko's adhesive force comes from the structure of the toepads on the tips of its toes.

Geckos are the largest animals that use adhesion to climb vertical walls.

Groups of sandcastle worms build their homes close together, forming large underwater reefs in the Pacific Ocean off the coast of California.

Gecko toepads are bristly with millions of tiny, super-flexible, hairlike parts called setae. Just as opposite poles of magnets attract and stick together, each setae develops an attractive force with the surface it touches. This happens because everything—including gecko setae—is made up of too-tiny-to-see building blocks called molecules. When two things touch, their surface molecules create a weak connection. This is like the attraction between the north and south poles of two magnets only much weaker. But the gecko has so many setae that the adhesive force is strong enough to hold the gecko in place. The little lizard stays stuck until it lifts its toes, peeling its setae off the surface.

A gecko can stick and unstick all the tiny setae on its feet fast enough that in only a second it can cross a distance equal to about twenty times its body length.

A sandcastle worm produces a gluelike substance to stick things together. It uses this to build its home. The substance is amazing because it stays stuck when it's wet, something most glues cannot do.

The home the sandcastle worm builds is more of a fort than a castle. It constructs this to protect itself from predators, such as fish and crabs. And it builds by picking up bits of shell or grains of sand one at a time with its tentacles. It passes each bit to pincerlike parts in front of its mouth. There, it dabs on a small amount of the adhesive liquid its body produces. Finally, the worm uses its tentacles to place the glue-coated bit of material onto the wall it's constructing. Once finished building, the worm lives inside its glued-together home. Besides staying safe, the worm keeps its body hidden while it pokes out just its tentacles to catch the tiny animals it eats.

So, how have these two different animals sparked bioinspiration?

This is a sandcastle worm without its protective castle.

People had tried for years and failed to mimic how a gecko climbs. Alfred Crosby, a scientist at the University of Massachusetts, began trying to solve that mystery in 1996. But it wasn't until 2000 that he made an important discovery. He said, "That was when I began looking at how [a gecko] having lots of little setae coming into contact at once with a surface affects the [adhesive] force that develops." But this knowledge wasn't enough to invent something useful.

As he continued his research, Crosby realized the key to creating a weight-bearing adhesive material

This Geckskin pad stuck to glass is supporting a 300-pound (136 kg) weight.

was to mimic the gecko's whole foot structure and not just its setae-covered toepads. The gecko's foot has tendons (stiff yet elastic cords) connecting its toe bones to its setae-covered toepad skin. That is unusual because typically tendons don't reach the skin but only connect an animal's bones to the specific muscles that move them. With each step, as a gecko's foot presses down on its toe bones, the tendons spread out the toepad skin, maximizing contact of the setae with the surface it is on. When the gecko lifts its foot, the tendons create an upward pressure to lift the toe, breaking the contact and the adhesion.

To mimic this tendon structure, Crosby and his team combined a stiff but flexible material (copying the foot tendons) with a soft material that drapes over a surface for maximum contact (copying the gecko's many setae). After lots of trial and error, the team discovered they could use either Kevlar or carbon fibers as the stiff but flexible backing and fuse it (through a secret process they developed) to a soft polyurethane or silicone. The resulting material could stick and hold an object to a surface and, when tugged, release it. They named this bioinspired material Geckskin.

The United States Defense Advanced Research Projects Agency, commonly referred to as DARPA, was inspired by Geckskin. But while Geckskin could have a very strong adhesive force, it had to be slowly peeled to break its hold. The agency wanted soldiers to be able to stick and unstick fast—the way geckos do. Then they could quickly scale the side of a building to surprise an enemy.

Elliot Hawkes, at the University of California, Santa Barbara, was one of the scientists who worked to create what DARPA wanted. He said, "The key was to evenly distribute the climber's [weight] load across adhesive hand pads." Each hand pad was made up of twenty-four postage stamp-sized rectangles. Each of those rectangles was made up of many silicone wedges bonded to fiberglass. This structure mimicked the gecko's setae by flexing and pressing against any surface to create an adhesive force. And each rectangle was attached to a spring, which acted like the gecko's tendons to press all the silicon wedges against a surface. That helped share the weight being supported across the entire hand pad. Then a tug let the spring also quickly peel the hand pad off the surface. A soldier could easily climb up a vertical wall—even the side of a glass skyscraper—thanks to these special hand pads.

Elliot Hawkes climbs a glass wall. The pads on his gloves provide the adhesive force and stick to the surface until he pulls his hand away. Cables from the adhesive pads connect to footrests that support his body weight.

Scientists observed this sandcastle worm building its tubelike home in an aquarium filled with water, clean sand, and white silica beads.

Russell Stewart, a researcher at the University of Utah, was fascinated by how a sandcastle worm's glue could stick even underwater. He realized if people could make a glue that stays stuck when it's wet, it would be perfect for delicate human surgeries. But to go from this bioinspired idea to mimicking what a sandcastle worm produces was a *BIG* challenge.

The breakthrough came when Stewart realized that something had to trigger the glue to harden after it was released into the wet environment. Through tests and rigorous analysis, he learned the trigger for a sandcastle worm's adhesive was exposure to seawater. Stewart and his team tested lots of chemical combinations for a bioinspired glue and possible triggers for hardening it. Stewart said, "What we created mimics the sandcastle worm glue in how it works." He and his team created a glue that combines two kinds of polymers (chains of similar molecules linked together) in a super-salty solution.

Like the two opposite ends of a magnet, one of the glue's polymers has a positive electrical charge and the other has a negative electrical charge. These two charged polymers are dissolved in a super-salty solution. That keeps them separate so the liquid remains sticky. But when this solution is injected into the human body, the saltiness spreads out into the less salty blood and tissues. Then the two polymers contact each other and link up. PRESTO! The sticky substance hardens. The result can be lifesaving, such as sealing an opening in a blood vessel to stop bleeding.

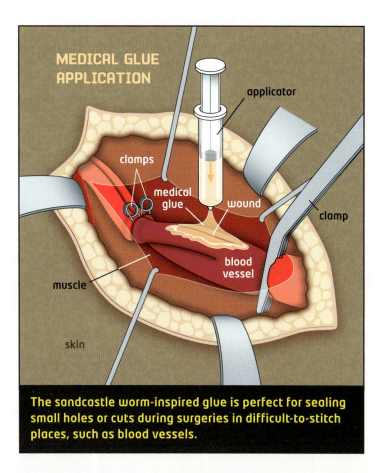

MEDICAL GLUE APPLICATION

applicator

clamps

medical glue

wound

clamp

blood vessel

muscle

skin

The sandcastle worm-inspired glue is perfect for sealing small holes or cuts during surgeries in difficult-to-stitch places, such as blood vessels.

DIVE MASTERS

Both diving beetles and water scorpions need to breathe air, and they both hunt for food underwater. So, they face the same challenge: *How to breathe underwater while they catch dinner.*

Both animals successfully adapted to tackle this challenge. But while they both live in freshwater—sometimes even share the same pond—how they adapted is very different.

Diving beetles carry an air bubble underwater with them to be their air supply while catching a fish, insect, or tadpole meal.

Water scorpions don't sting. They got their name from their long, taillike breathing tube.

A diving beetle captures the air it needs to breathe underwater by slapping its abdomen down on the water's surface as it dives. This motion creates air bubbles, and the beetle traps one between its shell-like wing cases (coverings over its folded wings) and its abdomen. The beetle's wing cases hold the air bubble over its spiracles (openings that let air into and out of the beetle's body), letting it breathe underwater. But that isn't the only oxygen the beetle gets during its dive. As the beetle uses up some of the oxygen in its bubble, oxygen in the water flows into the

The diving beetle can stay underwater for as long as fifteen minutes before it needs to surface and expose the spiracles on its abdomen to air.

bubble replacing what was lost. That happens because molecules of oxygen naturally move from areas of high concentration (where there is a lot) to areas of lower concentration (where there is less). But this adds oxygen to the air bubble so slowly that the diving beetle eventually uses up more oxygen than is replaced. Then it needs to surface to breathe and create a new bubble to dive again.

To breathe while it's underwater, a water scorpion has a special tail that's like a straw. It pokes the tip of its tail out of the water to breathe while the rest of its body stays underwater. The inside of the water scorpion's tail is lined with tiny hairs that trap air. These hairs capture enough air to allow the water scorpion to break its contact with the surface long enough for a quick dive. That helps it catch insects and small fish to eat.

So, how have these two different animals sparked bioinspiration?

Breathing with its tail above the surface, the water scorpion stays still until prey comes close enough to grab with its large, pincerlike front legs.

There is such a long history of people working underwater no one knows for sure what inspired some inventions that helped divers breathe. As early as the 1600s, people may have copied the diving beetle's method for carrying air underwater with them to salvage valuable items from sunken ships. They dropped a large bell-shaped container straight down into water. Like the diving beetle's abdomen slapping the water, the drop trapped an air bubble inside the bell. Once underwater with the diving bell, divers swam out to work and into it when they needed a breath. But unlike the diving beetle's air bubble, the air inside the bell didn't gain oxygen from the surrounding water. Instead,

the diving bell's air gradually became too full of carbon dioxide gas (the waste gas people breathe out). Then the divers would surface, and the diving bell would be hauled out of the water to allow the air inside to refresh.

In 1823 Charles and John Deane invented the diving helmet. Whether or not this was truly bioinspired by a water scorpion, it worked the same way. The diving helmet had a long hose connected to an air pump on board a boat. Air passed through the hose to the diver, and stale air was forced out of the helmet through a separate hose. With this invention, a diver could stay underwater for an extended period. But unlike the water scorpion,

Even a large diving bell could trap only a small amount of air.

Brothers Charles and John Deane got the idea for a diving helmet after they created a helmet for firefighters to breathe safely in a smoke-filled area.

This diver equipped with scuba gear breathes a regulated supply of air stored as compressed air in the backpack tank.

a human diver could not disconnect from the surface. So, a diver's movements were limited by how far the air hose would reach.

During World War II, another invention was created that mimicked the diving beetle's breathing method and let human divers spend an extended period underwater. This self-contained underwater breathing apparatus, nicknamed scuba, allowed swimming soldiers, called *frogmen*, to plant explosives on enemy ships. After the war, people used scuba gear to investigate ocean life and conduct work underwater, such as making ship repairs. When diving with scuba gear, people could also enjoy exploring underwater regions. But scuba gear allows divers to safely descend only to a depth of 130 feet (40 m). That's because water pressure gets greater the deeper a diver goes. Intense pressure makes it difficult and then impossible for a human's lungs to work properly.

From 1979 to 1987, Canadian ocean explorer and scientist René Théophile Nuytten worked to develop the Newtsuit. New technologies enabled him to improve on the earlier diving inventions. And the Newtsuit lets divers safely explore and perform tasks, such as pipeline repairs, at depths up to 900 feet (274 m). In addition to working tethered with air supplied through a hose from the surface, like wearing a diving helmet, a Newtsuit diver can work untethered. Then the suit's backpack breathing system supplies air for up to forty-eight hours.

This diver is wearing a Newtsuit. It is sealed to maintain surface level air pressure even at great depths.

ARMOR UP

Sharks live in the ocean, and pangolins live on land. But even though their environments are different, both face the same challenge: **How to protect their body.**

Sharks and pangolins have each adapted to solve this problem by developing a protective coating—their body armor. But a shark's armor is different from a pangolin's armor and so is the reason it needs protection.

A hammerhead shark's body is coated with tiny, overlapping denticles. These are regularly shed and replaced.

A pangolin is born with a set amount of scales. These get bigger as it grows.

The light-colored spots on the southern right whale and her calf are barnacles attached to their skin.

A shark needs to protect itself from barnacles attaching to its skin. Barnacles, cousins of lobsters and shrimp, anchor to ocean animals, such as whales, with a concrete-like glue. Each attached barnacle adds a little weight the animal must carry. The body of each barnacle also creates a bump that increases drag, requiring more effort to propel the animal through the water. That slows the animal's swimming speed.

To protect itself from barnacles, a shark's entire body is coated with denticles. Each denticle is like a little tooth because it has a soft inner pulp covered by hard enamel. Such a hard coat prevents barnacles from attaching. So, a shark doesn't have any extra weight to slow it down, and its denticles let water slip past its body easily. That helps a shark swim fast. A hammerhead shark, for example, regularly cruises at 15 miles (24 km) per hour and swims even faster in short bursts to catch the sardines and other fish it eats.

Denticle size is different for every kind of shark. This is a magnified image of a hammerhead shark's denticles. In fact, a hammerhead shark's denticles are so tiny that about ten thousand cover an area no bigger than a US penny.

A pangolin needs armor to protect itself from being bitten by predators. Most of its body is covered with scales made of keratin, the same tough material as human fingernails and a rhino's horn. Each scale is also covered by an extra-hard coating. That makes a pangolin's armor too tough for even a lion's powerful bite to puncture. If attacked, a pangolin rolls up to shield its unarmored body parts—its forehead, belly, and the insides of its legs.

So, how have these two different animals sparked bioinspiration?

Scales aren't a pangolin's only defense mechanism. Once curled up, it also produces stinky farts to drive its attacker away.

Since the first ships sailed the oceans, vessels have had a problem with biofouling. Biofouling is when living things, such as barnacles and algae, attach themselves to a surface, including a ship's hull. It adds weight and drag, which slows the vessel down. Any ship with biofouling is heavier and slower and requires more wind power or fuel to reach its destination. Biofouling can also keep a ship's rudder, propellers, and other equipment from working properly. For centuries the only way to deal with biofouling was to periodically haul the ship out of the water and scrape its hull clean. Later, people covered a ship's hull with tar, metal shields, or toxic chemical paints to repel barnacles and algae. But those coatings also poisoned any ocean animal that came close. Centuries passed and while concern grew over how the anti-biofouling methods were hurting ocean life, no one came up with a safe solution.

Then, in 2000, Anthony Brennan, a scientist at the University of Florida, had an idea—a bioinspired idea. He knew a shark's denticles kept its body free of biofouling. That meant coating ships with a material that mimicked a shark's body armor should protect them too. Brennan took impressions of shark denticles and examined them with a scanning electron microscope (a microscope that scans the surface of a specimen with an electron beam in place of light to create a super-magnified image of whatever is being studied). Those images showed a shark's denticles each had a repeating diamond pattern of ridges.

All the bumps on this sailboat's hull are barnacles.

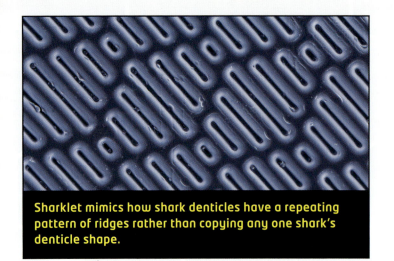

Sharklet mimics how shark denticles have a repeating pattern of ridges rather than copying any one shark's denticle shape.

It took several years of testing, failing, refining, and retesting to invent a successful coating that mimicked shark denticles. In 2007 Brennan's team produced a plastic film they named Sharklet because its surface mimicked a shark's denticle pattern. No barnacles or algae became attached to Sharklet-coated hulls. Further testing showed that Sharklet also prevented bacteria from sticking to a surface. So, it is being used on medical equipment instead of chemical disinfectants. And some people are considering using it as a coating for door handles and even backpacks.

At Golem Innovation in Bayonne, France, a research and development team created the Alpha helmet, which works like a pangolin's flexible armor. Weighing 17.6 ounces (500 g), it's a lightweight helmet made of tough panels. When not in use, these fold down to be a neck collar. The helmet gains extra shock-absorbing strength to protect the wearer against impact because, when unfolded, the panels interlock. Each interlocking connection becomes a reinforcement. That helps prevent cracking that could weaken the helmet's protective strength. Though the Alpha helmet is a good idea, the materials and the machinery needed to produce this bioinspired helmet make it too expensive for general use. But efforts to make it available continue.

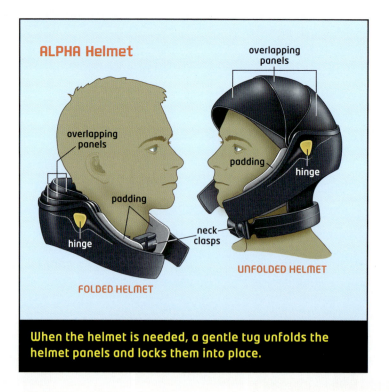

ALPHA Helmet

overlapping panels

overlapping panels

padding

hinge

padding

hinge

neck clasps

FOLDED HELMET

UNFOLDED HELMET

When the helmet is needed, a gentle tug unfolds the helmet panels and locks them into place.

WILD FUTURE!

You've discovered animal features and behaviors that have inspired inventions. But you have also learned these didn't happen simply because someone saw a cool animal and immediately created something useful. Most often a need came first—a need to be more comfortable, to be able to perform work more easily, or to be able to stay safe in a hazardous environment. When creative people discover a need, they start considering possible solutions. Then while searching for solutions, an animal with a special body feature or a specific behavior can spark an idea for an invention. When more than one animal adaptation offers ideas, usually where the invention needs to work and the available technologies and materials determines which becomes the winning bioinspiration.

So, remember, whenever you realize that something is needed to live more comfortably, stay safe, or make getting something done easier, look to animals for ideas.

Someday, you could be the problem solver who makes a difference with a *wild* invention.

Keep watching for how animals thrive, wherever you find them, thanks to their own unique adaptations for facing life's challenges.

A NOTE FROM SANDRA MARKLE

I have always been fascinated by how nature points the way for inventing something useful. After all, animals tackled the world's challenges first and those that were successful thrived. What surprised me as I researched *Wild Inventions* was how frequently two different animals faced a similar challenge—sometimes in completely different environments. Then each of those animals inspired people to create something useful—though often in different places or many years apart.

Writing *Wild Inventions* also made me realize it is essential to protect diverse animal species where they live in wild places. Who knows what challenges people will face and overcome in the future thanks to a yet-to-be-imagined, bioinspired invention?

GLOSSARY

architect: a person who designs buildings

biofouling: when living things, such as barnacles and algae, attach themselves to something

bioinspiration: designing and creating an invention after seeing how an animal has adapted to address a similar problem

camouflage: something that makes it possible to hide by blending into the surrounding area

echo: a reflected sound returning to a listener

echolocation: the ability to project sounds and evaluate the returning echoes to sense an environment

environment: the surroundings or conditions in which a person or animal lives

evaporation: liquid water turning into its gas form—water vapor—and using up heat during that process

habitat: an animal's natural home environment

larva: an immature form of some animals, such as honeybees, that are unlike the adult form

molecule: the smallest amount of any substance that has all its physical and chemical properties

predator: an animal that catches and eats other animals

prey: an animal eaten by a predator

saliva: watery liquid an animal secretes into its mouth

scuba: self-contained underwater breathing apparatus, or equipment that includes a tank of compressed gas, such as air, supplied to a swimmer at a regulated pressure through tubes and a mouthpiece. This allows a swimmer to explore or work underwater at depths up to 130 feet (40 m).

sonar: a system for detecting objects underwater by projecting sound pulses and recording the echoes after they've bounced off something

ultrasound: the diagnostic use of sound for looking inside the human body

SOURCE NOTES

11   Samuel Ocko, interview with the author, August 1, 2019.

12   Ocko.

18   Brian Branstetter, interview with the author, February 4, 2019.

19   Branstetter.

19   Laura Kloepper, interview with the author, February 1, 2019.

20   Kloepper.

28   Alfred Crosby, interview with the author, October 8, 2020.

29   Elliot Hawkes, interview with the author, September 28, 2021.

31   Russell Stewart, interview with the author, October 8, 2020.

DISCOVER MORE

Becker, Helaine. *Zoobots: Wild Robots Inspired by Real Animals.* Toronto: Kids Can, 2014.
Explore how roboticists are using unique abilities animals have as inspiration for designing robots capable of performing tasks humans can't do safely and easily or at all.

"Bio-Inspiration: Nature as Muse KQED Quest"
https://www.youtube.com/watch?v=JnBkbaFsZOY
A fun look at bioinspiration from the past to the present and a look to the future.

"Biomimicry: When Nature Inspires Design"
https://www.youtube.com/watch?v=ZODvr_GzNc4
See some unusual bioinspirations in action.

Nordstrom, Kristen. *Mimic Makers: Biomimicry Inventors Inspired by Nature.* Watertown, MA: Charlesbridge, 2021.
Dig into the text woven into illustrations to answer questions about how animals and plants inspired inventions.

"The Promise of Biomimcry"
https://www.youtube.com/watch?v=Muzfdq25Qbc
Biomimicry Institute offers a dynamic look at how the natural world has given us a chance to solve problems. What would nature do? The inventions and innovations inspired by nature are also sustainable and nondamaging to the environment.

Swanson, Jennifer. *Beastly Bionics: Rad Robots, Brilliant Biomimicry, and Incredible Inventions Inspired by Nature.* Washington, DC: National Geographic Kids, 2020.
Discover how the natural world is inspiring some amazing inventions.

INDEX

PHOTO ACKNOWLEDGMENTS

Image credits: Olha Pashkovska/iStockphoto/Getty Images, p. 4; Sergii Mikushev/Alamy Stock Photo, pp. 5; FLPA/Alamy Stock Photo, p. 7 (top); ultramarinfoto/iStock/Getty Images, p. 7 (bottom); Oxford Scientific/The Image Bank/Getty Images, pp. 8, 9; tomark/iStock/Getty Images, p. 10; Auscape/Universal Images Group/Getty Images, p. 11; Anna Henly Photography/Getty Images, p. 12 (left); Ginae McDonald/Shutterstock, p. 13; Thomas Cockrem/Alamy Stock Photo, p. 14; Ken Wilson-Max/Alamy Stock Photo, p. 15; RichardALock/E+/Getty Images, p. 16; DeAgostini/Getty Images, p. 17; Wildestanimal 2016/Getty Images, p. 18; Roy Riley/Alamy Stock Photo, p. 19; Laura Kloepper, Ecological Acoustics and Behavior Laboratory, p. 20; John Cancalosi/Alamy Stock Photo, p. 21; Jeff Overs/BBC News & Current Affairs/Getty Images, p. 22 (right); NatalyaBurova/iStock/Getty Images, p. 23; nicola fusco/Alamy Stock Photo, p. 24; Blue Planet Archive/Alamy Stock Photo, pp. 25, 40; Andrew Syred/Science Source, p. 26; Josh Silberg/Hakai Institute, p. 27; Image courtesy of Michael Bartlett and Daniel King, Alfred J. Crosby Research Group, University of Massachusetts Amherst, p. 28; Eric Eason/Biomimetics and Dextrous Manipulation Lab, Stanford, p. 29; blickwinkel/Alamy Stock Photo, pp. 32, 35; Nature Picture Library/Alamy Stock Photo, p. 33; Arterra Picture Library/Alamy Stock Photo, p. 34; Fox Photos/Hulton Archive/Getty Images, p. 36 (left); WaterFrame/Alamy Stock Photo, p. 36 (right); ultramarinfoto/E+/Getty Images, p. 37 (top); Pool DUCLOS/ROSSIGamma-Rapho/Getty Images, p. 37 (bottom); imageBROKER/Alamy Stock Photo, pp. 38, 39; Ted Kinsman/Science Source, p. 41; Frederick Mark Sheridan-Johnson/Alamy Stock Photo, p. 42; Simon McGill/Moment/Getty Images, p. 43; Pascal Goetgheluck/Science Source, p. 44 (left); Dmytro Zinkevych/Alamy Stock Photo, p. 45; Skip Jeffery Photography, p.46; Backgrounds: Nazhul/Shutterstock.

Cover: Byrdyak/iStock/Getty Images; Alexey_Seafarer/iStock/Getty Images; chanonttennis/Shutterstock; Nazhul/Shutterstock; Binturong-tonoscarpe/Shutterstock.